ISBN 978-1-5278-1876-7
PIBN 10894078

1 MONTH OF
FREE
READING

at
www.ForgottenBooks.com

By purchasing this book you are eligible for one month membership to ForgottenBooks.com, giving you unlimited access to our entire collection of over 1,000,000 titles via our web site and mobile apps.

To claim your free month visit: www.forgottenbooks.com/free894078

English
Français
Deutsche
Italiano
Español
Português

www.forgottenbooks.com

Mythology Photography **Fiction**
Fishing Christianity **Art** Cooking
Essays Buddhism Freemasonry
Medicine **Biology** Music **Ancient**
Egypt Evolution Carpentry Physics
Dance Geology **Mathematics** Fitness
Shakespeare **Folklore** Yoga Marketing
Confidence Immortality Biographies
Poetry **Psychology** Witchcraft
Electronics Chemistry History **Law**
Accounting **Philosophy** Anthropology
Alchemy Drama Quantum Mechanics
Atheism Sexual Health **Ancient History**
Entrepreneurship Languages Sport
Paleontology Needlework Islam
Metaphysics Investment Archaeology
Parenting Statistics Criminology
Motivational

e Document

eflects current scientific knowledge,

MASS SOIL MOVEMENTS
IN THE H.J. ANDREWS
EXPERIMENTAL FOREST

C. T. DYRNESS

PACIFIC NORTHWEST
FOREST AND RANGE EXPERIMENT STATION
U.S. DEPARTMENT OF AGRICULTURE
U.S. FOREST SERVICE Research Paper PNW-42
1967

CONTENTS

INTRODUCTION

Each winter a number of mass soil movements, such as earthflows and slumps, occur in the Coast Ranges and Cascades of western Oregon. These movements, generally occurring during periods of heavy and prolonged rainfall, often disrupt roads and damage other improvements. Some, however, remain undiscovered in remote, undisturbed areas.

During the winter of 1964-65, there was an unusually large number of mass soil movements in western Oregon. A high proportion of these movements occurred during the severe storm that struck northern California and Oregon the week before Christmas. Soon after this storm, it was found that the H. J. Andrews Experimental Forest was in one of the hardest hit areas and was in an excellent location for an intensive survey of storm damage with emphasis on the origin of large mass soil movements. Accordingly, a survey of all mass movements in the Experimental Forest was initiated April 1 and concluded July 1, 1965.

The study had several aims: (1) to completely describe and photograph all major movement areas in the Andrews so that changes in the future may be followed and documented, (2) to identify the types of movement which had occurred and estimate the amount of damage to the site and/or improvements, (3) to attempt to define the relationship between the mass soil movement and (a) amount of man's disturbance and (b) certain site factors such as geology, soil, elevation, and aspect.

STUDY AREA

The H. J. Andrews Experimental Forest[1] comprises the entire 15,000-acre drainage of Lookout Creek, located approximately 40 miles east of Eugene, Oreg., and part of the Willamette National Forest. The area lies within the Western Cascades physiographic province and is predominantly mature topography with sharp ridges and steep slopes. Bedrock in the area is made up of pyroclastics (tuffs and breccias) and basic igneous rocks (basalt and andesite). Forest cover is comprised of old-growth Douglas-fir *(Pseudotsuga menziesii)*, with varying amounts of western hemlock *(Tsuga heterophylla)* and western redcedar *(Thuja plicata)*.

THE STORM

Fredriksen has described the characteristics of the 1964 Christmas storm in the H. J. Andrews area in some detail.[2] At lower elevations, more than 13 inches of rain in 4 days was supplemented by an undetermined amount of snowmelt. This storm has generally been classified as an event with a 50-year return period. However, Fredriksen pointed out that in

---

[1] Berntsen, Carl, M., and Rothacher, Jack. A guide to the H. J. Andrews Experimental Forest. Pacific Northwest Forest & Range Exp. Sta., 21 pp., illus. 1959.

[2] Fredriksen, R. L. Christmas storm damage on the H. J. Andrews Experimental Forest. U.S. Forest Serv. Res. Note PNW-29, 11 pp., illus. 1965.

only 12 years of record at the Experimental Forest, this was the second storm to total more than 13 inches of precipitation within a 4-day period. He concluded, "we believe 4-day storms which deliver 13 inches are probably not unusual." The fact remains, however, that this storm caused the most extensive damage recorded since establishment of the Experimental Forest in 1948.

METHODS

Information recorded at each soil movement site included type and specific characteristics of soil or debris movement, general characteristics of the area, and assessment of factors influencing the soil or debris movement. Photographs were also taken at each site and a sketch made showing the outstanding features. An attempt was made to find and describe all events where 100 cubic yards or more of soil or debris were involved.

RESULTS

Descriptions of 47 mass movement events were made in the Andrews Forest (fig. 1). Estimated sizes ranged from 100 to 75,000 cubic yards, and the total quantity of material moved was about 347,800 cubic yards.

Mass movements
by morphological class

The movements have been grouped into 10 classes on the basis of their morphological characteristics (table 1). Earthflows occur when the soil becomes saturated, and pore-water pressure builds up to such an extent that the soil mass flows out in a generally tongue-shaped form. Slumps, also, generally occur in saturated soils; however, they are characterized by rotational movement and spoon-shaped failure planes.

Table 1.--Mass movements occurring on the H. J. Andrews Experimental Forest during the winter of 1964-65, by morphological class

Morphological class	Events	Material moved
	Number	Cubic yards
Earthflow	18	104,600
Slump with earthflow	7	14,525
Slump	3	4,050
Earthflow causing channel scour	5	82,350
Slump causing channel scour	4	106,300
Debris slide causing channel scour	2	7,500
Channel scour	3	26,300
Roadfill washout	3	1,050
Avalanche with rilling	1	100
Debris avalanche with earthflow	1	1,000
Total	47	347,775

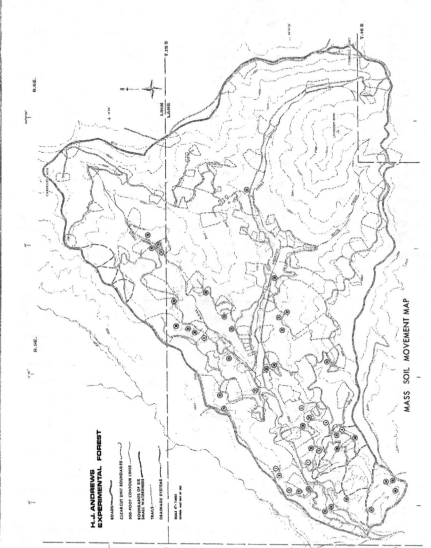

Figure 1.—Map of the H. J. Andrews Experimental Forest showing location of mass movement events.

Channel scouring is generally caused by massive and sudden failure of the entire unconsolidated mantle, most often in areas of the oversteepened drainage heads. Debris avalanches involve the sudden downslope movement of largely rock material in steep terrain. The earthflow, slump, and avalanche classes are defined by Eckel.[3/]

Earthflows were the most common class of soil movement with 18 out of the total 47 events falling within this class (table 1). Channel scouring events ranked second with a total of 14. Only three of these events were confined entirely to the drainage channel; the remainder were triggered by a separate mass movement which frequently began at a road. The third most commonly occurring movement class was the slump with some earthflow characteristics. Soil material which has slumped down, when saturated with water, will frequently flow out at the toe and may travel a great distance downslope. Only three slump events were completely devoid of flow characteristics. These were backslope slumps where the material came to rest on the road surface.

A consideration of amounts of material moved emphasizes the importance of earthflow and channel scouring events (table 1). The remaining forms of movement contributed very little to the total amount of material moved.

Mass movement by mode of action or disturbance

Mass movements on the Andrews were also classified according to type of disturbance or mode of action (table 2), with attention focused on the source area of the event. For

Table 2.--Mass movements occurring on the H. J. Andrews Experimental Forest during the winter of 1964-65, by mode of action or disturbance

Type of disturbance	Events	Material moved
	Number	Cubic yards
Roadfill failure	12	40,900
Road backslope failure	5	14,400
Road backslope and fill failure	6	38,325
Events caused by road drainage water	8	86,450
Road removed by stream	3	4,350
Events in logged areas	8	22,150
Events in undisturbed areas	5	141,200
Total	47	347,775

[3/] Eckel, Edwin B., ed. Landslides and engineering practice. Highway Res. Board Spec. Rep. 29, 232 pp., illus. Washington, D.C. 1958.

example, when a channel scouring event originated in an untouched stand of timber, it was classified as a movement occurring in an undisturbed area even though at a downstream location, roads may have been severely damaged.

Although only five events occurred in undisturbed areas, this type contained the largest volume of material. Events caused by road drainage water ranked second, with the other five types far behind. Each class is considered separately below:

1. Roadfill failure.

The most common single type of mass movement on the Andrews was the roadfill failure (12 events) (fig. 2). This is apparently a rather simple type of movement caused by the saturation of the roadfill embankment. Most of the fill materials, especially those derived from pyroclastics, appear to flow readily when saturated. As nearly as could be determined, these failures were not due to improper road drainage but, rather, were caused by the unusually large amounts of rainfall or perhaps by improper fill compaction.

2. Road backslope failure.

This is a common event, which frequently results in road blockage (fig. 3). The failure most often takes the form of a slump, but the slump material may have some earthflow characteristics near the toe. The most failure-prone backslopes were those constructed in areas of deep, well-weathered tuffs and breccias. These saprolitic materials fail easily when excessively moist. If failures did occur in areas of andesite or basalt parent material, invariably the local bedrock was highly fractured. Like the majority of other mass movement events, backslope slumps occur only when the mantle is saturated and are also more common in areas where mass movement has occurred repeatedly in the past.

Figure 3.--Backslope failure in an area of soil derived from andesite colluvium.

Figure 2.--Roadfill failure in the H. J. Andrews Experimental Forest.

5

3. Road backslope slump causing fill failure.

Six events in the Andrews fit within this type (fig. 4). In effect, this is a "chain reaction" with the general sequence as follows: A portion of the backslope becomes saturated and slumps down onto the road surface, thus blocking the inside drainage ditch. The drainage water is then diverted across the road surface and down onto the fill slope from the outside portion of the road. The fill slope becomes saturated with water and an earthflow type of movement follows, encroaching to a varying extent onto the road surface. In some cases, the "chain" is further lengthened by the fact that the earthflow on the fill slope may trigger channel scouring in a drainage below.

4. Mass movement caused by concentration of road drainage water.

There are two classes of events under this heading: (a) those due to the concentration of road drainage water, even though the road drainage system was apparently functioning properly (fig. 5), and (b) those caused by failure of some portion of the drainage system--for example, a clogged culvert or inside ditch.

Events caused by concentration of road drainage water. --Four events in the Andrews fall within this category. In all cases, the event was associated with stream channel scouring. The typical sequence was as follows: Extremely large quantities of water flowing from a culvert completely saturated the soil mantle below the road. In most cases, this saturated area was

Figure 4.--View of typical backslope slump which has also caused roadfill failure.

near the head of a tributary drainage. When the soil became saturated, the entire mantle either slumped or flowed downslope and temporarily blocked the drainage channel. When a sufficient head of water was built up, this temporary dam breached, and a wall of water and debris rushed down the channel removing everything in its path. The end result was a drainage channel from which virtually all unconsolidated material has been removed exposing bare rock.

All four events of this type occurred in areas of hard, fractured, greenish breccias and tuffs.

Figure 5.--Soil movement at the head of a drainage caused by high flows through the culvert and inadequate energy dissipation at the outlet end.

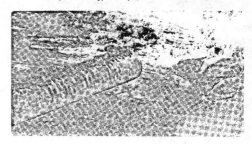

6

Events caused by failures in the road drainage system. --Of four of these events in the Andrews, three were 200 to 500 cubic yards in size, while the fourth involved approximately 75,000 cubic yards of material. In every case, these failures occurred in areas of deeply weathered tuffs and breccias. All were flow-or slump-type movements of roadfill or sidecast material, none of which resulted in channel scouring. In these cases, the concentration of water resulting in the necessary saturation came from the clogging of a culvert or inside ditch (fig. 6). These blockages had grave consequences because of the tremendous volumes of water diverted onto the roadfill or sidecast materials.

Figure 6.--Massive roadfill failure caused by blockage of road drainage ditch.

5. Road removed or damaged by stream.

   In three cases, culverts carrying perennial or intermittent streams failed either because of inadequate size or, more probably, because they became blocked by debris (fig. 7). As a result, roads were completely washed out or were severely damaged.

Figure 7.--Debris which blocked the culvert and caused several feet of material to be deposited on the road surface. (Road, in foreground, has been repaired.)

6. Mass movement events in logged areas.

The eight mass movement events in logged-over areas were principally earthflows (fig. 8), although two also had some slump characteristics. They ranged in size from 150 to 9,500 cubic yards of material. Four of the movements occurred in areas which gave ample evidence of considerable mass soil movement in the past (i.e., very uneven, hummocky relief). The remaining four were on smooth, steep slopes. In two cases, where the movements occurred at the heads of drainages, there was extensive channel scouring.

Although these movements took place in disturbed, logged areas, in every case it is impossible to say whether the logging per se caused the instability. It is probably safe to assume that at least one or two of them would have occurred even if the area had not been logged.

7. Mass movement events in undisturbed areas.

Of five mass movement events in completely undisturbed areas, all were connected with channel scouring (fig. 9), but three also had earthflow characteristics at the source area. Size of these movements ranged from 200 to 75,000 cubic yards--four were among the largest slides that occurred on the Andrews. The average for all five events was well over 28,000 cubic yards.

Figure 9.--Channel scour event in undisturbed timber.

With one exception, these events constituted failures at the heads of drainages. These areas are often very steep and unstable due to the headward erosion of the stream channel. When these areas become completely saturated, the weight of water adds to the driving force and at the same time decreases the strength of the soil. Sudden, massive failures frequently result. This is, undoubtedly, the normal course of geologic erosion in this area, and there is little that man can do to forestall events of this type.

Relationship between mass movements and certain site characteristics

The influence of roads on mass movements in the Andrews is clearly indicated in table 3. Although over 72 percent of the mass movements occurred in connection with roads, only 1.8 percent of the total area of the forest is in road rights-of-way. On the other hand, only about 11 percent of the events occurred in undisturbed areas. Since 84.6 percent of the total area is undisturbed, this is far less than would be expected if the events occurred in a random manner.

In the western Cascades, it has been observed that mass soil movements occur much more frequently in areas of pyroclastic rocks (tuffs and breccias) than in areas where the bedrock is comprised of basalt or andesite. It has also been noted elsewhere that greenish tuffs and breccias are more unstable than are their reddish counterparts. These relationships are borne out by the data presented in table 3. About 94 percent of the events occurred in areas with a tuff and/or breccia substratum, even though only 37 percent of the total area is made up of these rocks. Moreover, 64 percent of the mass movements were on greenish tuffs and breccias which make up only 8 percent of the total area--clearly indicating the unstable nature of these materials. Although about 63 percent of the area is underlain by basalt and andesite materials, only 6.4 percent of the mass movement events occurred there.

When soil series in the movement area are considered, the relationship described above still holds but to a lesser extent. Eighty-seven percent of the events occurred in areas of soil from tuffs and breccias-- the breakdown being almost 49 percent on soils from reddish tuffs and breccias and 38 percent on soils from greenish tuffs and breccias. The apparent discrepancy between these figures and those for the substratum reported above may be attributed to the fact that many soils have been formed in transported parent materials and often there is little relationship between the soil profile and the substratum. It was noted in several cases that the discontinuity between the soil profile and the underlying bedrock constituted the failure plane, probably due to inherent weakness and water flowing through the zone.

Elevational range in the Andrews Experimental Forest is approximately 1,400-5,300 feet. All but two of the mass movements occurred at elevations below 2,900 feet (fig. 10). The great bulk of the events (over 72 percent) occurred in the elevational range from 2,000 to 2,600 feet (table 3). There are two possible reasons for the lack of mass movements at high elevations: (1) Most of the deposits of tuffs and breccias occur at

Table 3.--Relationship between occurrence of mass movement events and certain site factors

in the H. J. Andrews Experimental Forest during winter of 1964-65

Site factors	Mass movement events	Total area	Events per 1,000 acres
	----------- Percent -----------		---- Number ----
Disturbance:			
Undisturbed	10.6	84.6	0.4
Logging	17.0	13.6	3.9
Road construction	72.4	1.8	125.9
Substratum in movement area:			
Greenish tuffs and breccias	63.8	8.0	25.0
Reddish tuffs and breccias	29.8	29.2	3.2
Andesite colluvium	4.3	42.7	.3
Basalt and andesite residuum	2.1	20.1	.3
Soil series:			
Soils from reddish tuffs and breccias	48.9	22.8	14.2
Frissel Series	27.6	9.3	9.3
McKenzie River Series	21.3	13.5	4.9
Soils from greenish tuffs and breccias	38.3	9.1	37.2
Limberlost Series	25.5	4.1	19.5
Budworm Series	4.3	3.0	4.4
Slipout Series	8.5	2.0	13.3
Soil from andesite colluvium			
Carpenter Series	10.6	34.9	.9
Rock land	2.1	4.1	1.6
Elevation (feet):			
1,400-1,700	2.1	2.7	2.5
1,700-2,000	8.5	6.2	4.3
2,000-2,300	31.9	8.8	11.4
2,300-2,600	40.4	10.9	11.6
2,600-2,900	12.8	13.1	3.0
2,900-3,200	2.1	10.9	.6
Over 3,200	2.1	47.4	.1
Slope (percent):			
0-15	2.1	5.9	1.1
15-30	0	25.2	0
30-45	14.9	34.6	1.3
45-60	36.2	22.8	5.0
60-75	31.9	8.4	11.9
75-90	14.9	2.5	18.7
Over 90	0	.6	0
Aspect:			
North	8.5	16.8	1.6
Northeast	14.9	5.0	9.3
East	8.5	5.9	4.5
Southeast	17.0	11.4	4.7
South	0	11.9	0
Southwest	4.3	16.8	.8
West	19.2	9.9	6.1
Northwest	25.5	19.3	4.1
Level	2.1	3.0	2.2

ower elevations; (2) temperatures prevailing above 3,000 feet at the time of the December storm may have been low enough to maintain some snow cover.

As would be expected, steepness of slope and frequency of slides were apparently positively correlated (table 3). Only about a sixth of the vents occurred on slopes with a gradient of less than 45 percent (fig. 10).

For some reason, mass movements were almost nonexistent on slopes with a south or southwest aspect-- probably because rock weathering and soil formation proceeds much more slowly on the drier aspects. The resulting shallow soils and less deeply weathered rocks may give rise to a greater degree of stability. It appears that mass soil movements occur more readily in areas of deep soils and well-weathered bedrock.

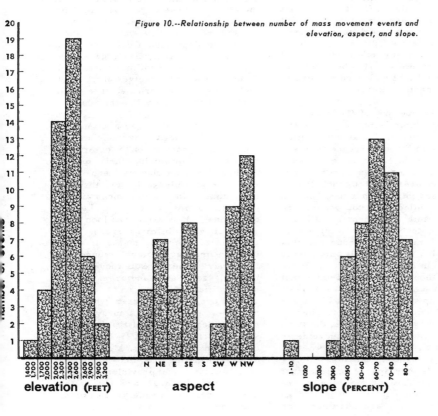

Figure 10.--Relationship between number of mass movement events and elevation, aspect, and slope.

DISCUSSION AND CONCLUSIONS

Mass soil movements are the end result of a combination of causative factors. In our area, these generally include soil materials of low internal strength, sufficient water for saturation or near saturation, and at least moderately steep slopes. Man's activities may contribute to movement by changing the slope or other characteristics of the mantle and by influencing the distribution of water. Since a complex of factors interact in the production of mass movements, it is often extremely difficult to pinpoint the specific cause of a mass movement event. In a very real sense, every movement event differs in some respect from any other event, and each represents a particular combination of conditions which culminates in massive mantle failure.

Perhaps one of the most important problems facing us is to determine the extent that man's activities contribute to the occurrence of mass soil movements. From our experience in the H. J. Andrews Experimental Forest, we are forced to conclude that man-caused disturbance has had an important influence in the occurrence of mass soil movements. Although logging disturbance also apparently increased movement frequency, this relationship is especially marked in the case of road construction. However, it should be borne in mind that in an area such as the Andrews Forest, where slopes are steep and a large portion is underlain by soft, deeply weathered pyroclastic rocks, the stage is set for extensive mass movements during high rainfall periods whether or not disturbance is a factor. Therefore, it is perhaps often true that man's activities accelerate the occurrence of mantle failure events in an already unstable area, rather than

contribute in any significant way to this basic instability. In other words, disturbance may cause some small and, by itself, insignificant change which is nonetheless sufficient to upset the tenuous equilibrium and trigger mass soil movement.

It is perhaps of some significance that the largest mass movement events encountered in this study were those which occurred in undisturbed areas. All involved extensive stream channel scouring, and the source areas were oversteepened drainage heads. This general pattern is interpreted as the normal course of geologic erosion. On the other hand, man-caused movements are generally smaller, since the entire slope--from drainage head to main stream channel--is generally not involved. Instead, many events in disturbed areas involve only road rights-of-way, or perhaps only a small portion of the slope below the road.

Since roads are so often an important factor in causing mass movements, the problem is to determine means of minimizing their effect. Perhaps the most obvious means is to reduce road mileage to an absolute minimum. In steep, mountainous terrain, this may be done by the use of skyline and, possibly, balloon logging methods. In many areas, it is possible that improvements in road location may appreciably reduce the frequency of mass soil movements. Unstable soils and landforms should be identified, and the route selected should avoid these areas wherever possible. In addition, improvements in road design and construction may also contribute substantially to increased mantle stability. Modification of waste handling to avoid sidecasting on steep slopes and provision for adequate road drainage are two of the more important means to minimize mass movement hazard.

12

Dyrness, C. T.
1967.   Mass soil movements in the H. J. Andrews
        Experimental Forest. U.S. Forest Serv.
        Res. Pap. PNW-42, 12 pp., illus. Pacific
        Northwest Forest & Range Experiment Sta-
        tion, Portland, Oreg.

    Analyzes 47 mass movement events resulting from
severe storms during the winter of 1964-65. Earthflow
and channel scouring events were the most common.
About 72 percent of the mass movements occurred in
connection with roads and 17 percent in logged areas.
Over 94 percent of the events occurred in areas of tuff
and/or breccia bedrock which occupy only 37 percent of
the total area.

Dyrness, C. T.
1967.   Mass soil movements in the H. J. Andrews
        Experimental Forest. U.S. Forest Serv.
        Res. Pap. PNW-42, 12 pp., illus. Pacific
        Northwest Forest & Range Experiment Sta-
        tion, Portland, Oreg.

    Analyzes 47 mass movement events resulting from
severe storms during the winter of 1964-65. Earthflow
and channel scouring events were the most common.
About 72 percent of the mass movements occurred in
connection with roads and 17 percent in logged areas.
Over 94 percent of the events occurred in areas of tuff
and/or breccia bedrock which occupy only 37 percent of
the total area.

Dyrness, C. T.
1967.   Mass soil movements in the H. J. Andrews
        Experimental Forest. U.S. Forest Serv.
        Res. Pap. PNW-42, 12 pp., illus. Pacific
        Northwest Forest & Range Experiment Sta-
        tion, Portland, Oreg.

    Analyzes 47 mass movement events resulting from
severe storms during the winter of 1964-65. Earthflow
and channel scouring events were the most common.
About 72 percent of the mass movements occurred in
connection with roads and 17 percent in logged areas.
Over 94 percent of the events occurred in areas of tuff
and/or breccia bedrock which occupy only 37 percent of
the total area.

Dyrness, C. T.
1967.   Mass soil movements in the H. J. Andrews
        Experimental Forest. U.S. Forest Serv.
        Res. Pap. PNW-42, 12 pp., illus. Pacific
        Northwest Forest & Range Experiment Sta-
        tion, Portland, Oreg.

    Analyzes 47 mass movement events resulting from
severe storms during the winter of 1964-65. Earthflow
and channel scouring events were the most common.
About 72 percent of the mass movements occurred in
connection with roads and 17 percent in logged areas.
Over 94 percent of the events occurred in areas of tuff
and/ o breccia bedrock which occupy only 37 percent of
the total area.

The FOREST SERVICE of the
U. S. DEPARTMENT OF AGRICULTURE
is dedicated to the principle of mul-
tiple use management of the Nation's
forest resources for sustained yields
of wood, water, forage, wildlife, and
recreation. Through forestry research,
cooperation with the States and private
forest owners, and management of
the National Forests and National
Grasslands, it strives — as directed
by Congress — to provide increasingly
greater service to a growing Nation.

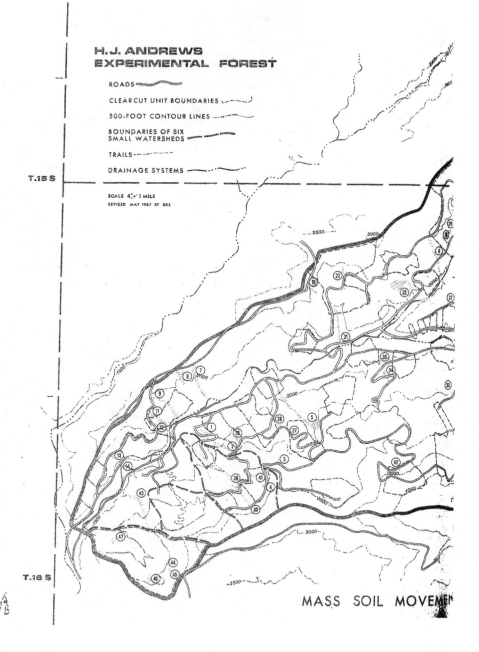

H.J. ANDREWS
EXPERIMENTAL FOREST

ROADS

CLEARCUT UNIT BOUNDARIES

500-FOOT CONTOUR LINES

BOUNDARIES OF SIX
SMALL WATERSHEDS

TRAILS

DRAINAGE SYSTEMS

SCALE 4" = 1 MILE
REVISED MAY 1967 BY BKS

T.15 S

T.16 S

MASS SOIL MOVEMEN